I0049203

MARINE RAIDERS

FITNESS PREPARATION LOG

INTRODUCTION

The Marine Corps Forces Special Operations Command (MARSOC) is preparing professional combat athletes who are ready to execute missions in the largely unstructured battlefields of today and the future. MARSOC's performance and resilience ideology—which focuses on mind, body, and spirit—is integral to preparation and success.

MARSOC seeks to ensure that Marines are not only physically strong but also possess the mental focus and unconquerable spirit necessary to persevere under the extreme stress of a high operational tempo and through the unknowns of asymmetric warfare. Critical skills operators are trained with renewal in mind, focusing on the resilience of the individual, unit, and family.

This fitness program focuses on improving physical performance through exercise and nutrition. It provides photographs and descriptions of exercises used at MARSOC, and is designed to prepare candidates for the physical aspects of Assessment and Selection (A&S). Upon arriving at A&S, candidates are expected to have completed this 10-week program.

The Three Pillars

In addition to preparing candidates for the rigors of Phases I and II of A&S, this 10-week program prepares candidates for the following three fitness tests:

The MARSOC In Test

The United States Marine Corps Modified Intermediate Level Swim Test

The United States Marine Corps Special Operations Command Assessment & Selection Ruck Movement Standards

The schedule is developed to give each event—running, swimming, and rucking—at least two training days a week.

Some workout days have MARSOC In Test preparation sections that are meant specifically to address pull-ups or crunches. Ruck workouts on Tuesdays incorporate exercises that will help you increase your speed with a ruck, as well as develop your ability to carry it during hard and physically demanding movements.

The fitness program outlines the maximum amount of daily events to be conducted. Individuals should strive to complete all of the daily events for the best results possible. Individual training goals should take into consideration levels of fitness in each of the core areas and focus on areas of weakness while increasing overall strength.

CONTENTS

Movement Preparation

A warm-up that prepares your body for movement, training, and performance. It boosts your heart rate, increases blood flow to the muscles, and elevates your core temperature.

Calisthenics

Exercises designed to develop muscular tone and promote physical well-being, relying heavily on body weight with minimal equipment requirements.

CONTENTS

Post-Workout Regeneration

Activities that increase the body's ability to recover faster, in order to maximize the gains achieved through performance training.

Nutrition, Hydration, Foot Care, and Recovery

Guidelines that help you select the right foods and beverages for optimum physical performance.

MOVEMENT PREPARATION

Ankles-Hips-Shoulders

Starting Position: Left knee bent at 90 degrees, with foot flat on the ground. Right knee is on the ground with foot flexed and arms down at your side.

1) Lean forward slightly and raise hands to back of head, with palms facing forward.

2) Twist to the left, pause, twist back to the front.

Do 5 reps, then switch legs.

Hip Bridge

Starting Position: Lying on back with arms away from your sides, with feet flexed.

1) Raise your hips off the ground, forming an arch, resting your weight on your shoulder blades (not your neck). Use your hands for balance. Hold for 15 seconds and return to starting position.

Do 6 reps.

Elbow Push-Ups

Starting Position: Start in plank position, lying horizontal on forearms, fingers and thumbs together with hands just in front of your face. Feet flexed.

1) Push up through your forearms, keeping back straight until your body is off the ground. Pause at the top.

2) Keeping back straight, return to starting position, just above the ground. Do not rest on ground.

Do 6-8 reps.

Bird Dog

Starting Position: Kneeling with palms and knees on the ground, feet flexed; knees directly below hips.

1) Lift and extend left arm and right leg simultaneously. Body and extended arms should be parallel to the ground. Hold for 30 seconds.

2) Return to starting position.

3) Repeat exercise while switching arms and legs that are extended.

Do 5 reps.

Frog Squats

Starting Position: Standing with feet more than shoulder width apart.

1) Squat as deep as possible with hips back and heels on the ground, hands together at the same level as your knees.

2) Return to starting position.

Do 5 reps.

World's Greatest Stretch

Starting Position: Standing with feet shoulder width apart, hands at your sides.

1) Step forward with your left foot, into a lunge position. Bringing your left elbow to your left knee, reach across to grab your right bicep.

2) Twist your body and rotate up and to the left, reaching up with your left hand.

3) Bring your hands back down, returning to the lunge position and then back to the starting position.

Do 5 reps, then switch sides.

MOVEMENT PREPARATION

Inchworm

Starting Position: Face down with arms fully extended and hands and toes on the ground.

1) Slowly walk your legs forward while keeping your hands in place, flat on the ground.

2) Go as far as you can with hands on the ground, then return to the starting position.

Do 5 reps.

Walking High Knees

Starting Position: Standing with feet shoulder width apart, hands at your sides.

1) Sharply raise left leg level with hips, bending at the knee so that thigh and foot are parallel with the ground, simultaneously pumping right arm.

2) Keep back straight and lower left leg and right arm while raising opposite arm and leg.

Do 10 reps.

Walking Quads Pulls

Starting Position: Standing with feet shoulder width apart, hands at your sides.

1) Sharply raise lower left leg up and back, stretching quadriceps. Flex foot, bring right arm up for balance.

2) Keep back straight and lower left leg and right arm while repeating step one with opposite arm and leg.

Do 10 reps.

Cradles

Starting Position: Standing with feet shoulder width apart.

1) Raise right leg in front of left leg and grasp right calf with both hands so that calf is parallel to the ground. Hold and balance for 30 seconds.

2) Return to starting position.

Switch legs and repeat 5 times.

Backwards Hamstring

Starting Position: Standing with feet shoulder width apart, hands at your sides.

1) In one movement, bend forward, with both arms forward, and simultaneously bring one leg up, fully extended.

2) Balance on one leg, with foot flat on the ground. Hold for 30 seconds and return to starting position.

Repeat on opposite leg. Do 5 reps.

High Knees

Starting Position: Standing with feet shoulder width apart, hands at your sides.

1) Bring the left knee up as high as possible, swinging the right arm up to cheek level.

2) Lower leg and repeat with opposite side.

Repeat for 15 seconds.

Butt Kicks

Starting Position: Standing with feet shoulder width apart, hands at your sides.

1) Standing straight, lift left foot up and back so that heel touches buttocks.

2) Return to starting position and alternate legs. Alternate pumping opposite arms for balance.

Repeat for 15 seconds.

MOVEMENT PREPARATION

For all exercises that begin with one leg, begin with the
LEFT LEG FIRST. For most exercises, the instructor counts
cadence and the student counts the number of reps. During
some exercises, the instructor does all of the counting.

MOVEMENT PREP CARD

Movement (page #)	Reps/time
1 **Ankles-Hips-Shoulders** (6)	5 each side (5 count)
2 **Hip Bridge** (6)	6
3 **Elbow Push-Ups** (6)	6-8
4 **Bird Dog** (7)	5 each side
5 **Frog Squats** (7)	5
6 **World's Greatest Stretch** (7)	5 each side
7 **Inchworm** (8)	5
8 **Walking High Knees** (8)	10 each side
9 **Walking Quads** (8)	10 each side
10 **Cradles** (9)	5 each side
11 **Backwards Hamstring** (9)	5 each side
12 **High Knees** (9)	15 sec
13 **Butt Kicks** (9)	15 sec

Forward Plank

Starting Position: Lying on stomach, with elbows bent and legs extended, forearms on the ground, and feet flexed.

1) Stiffen your core, lifting your body up in one movement so that you are resting on your forearms and toes. Your body should be straight.

Side Plank

Starting Position: On your left side with your body in a straight line (head, shoulder, hip, knees and ankles should be aligned) and forearm underneath you.

1) Push up on your left forearm, lifting your right hip, creating a straight line from your shoulder to your ankle. Only your left foot and left forearm should remain on the ground. Do not sag or bend at the waist. To increase difficulty raise your top leg to the sky.

Flutter Kicks

Starting Position: Lying on back with hands tucked under tailbone.

1) Raise your legs together 6-12 inches.

2) Alternate moving legs up and down in a scissor motion.

Push-Ups

Starting Position: Lying prone with feet flexed, elbows bent and hands under shoulders.

1) Push up through your hands, keeping back straight until elbows are fully extended. Pause at the top.

2) Keeping back straight, return to starting position, just above the ground. Do not rest on ground.

Lunges with Counter Rotation

Starting Position: Standing with feet shoulder width apart, with elbows out to the side and hands near your ears.

1) Step left leg forward into a lunge position, with knee over your toes. Right leg should be back, with knee pointed toward the ground and foot flexed.

2) Rotate your body toward your forward knee, keeping your spine straight.

3) Come back to center.

4) Drive off the left leg and return to standing position.

Frog Squats

Starting Position: Standing with feet greater than shoulder width apart.

1) Squat as deep as possible with hips back and heels on the ground, hands together at the same level as your knees.

2) Return to starting position.

Ammo Can Push Press

Starting Position: Keeping a neutral neck and flat back, and standing with feet just wider than shoulder width and toes slightly pointed out, grasp a 30 lb ammo can (or dumbbell) and rest it lightly on top of your upper chest.

1) Dip body by bending knees, hips and ankles slightly.

2) Explosively drive upward with legs, driving ammo can off chest by vigorously extending arms overhead.

3) Return ammo can to front of upper chest and repeat.

Ammo Can Thruster

Starting Position: Keeping a neutral neck and flat back, and standing with feet just wider than shoulder width and toes slightly pointed out, grasp a 30 lb ammo can (or dumbbell) and rest it lightly on top of your upper chest.

1) Descend into a squat by pushing your hips and butt back while inhaling. Continue down until the upper legs are at or just below parallel to the floor.

2) Descend into a squat by pushing your hips and butt back. Continue down until the upper legs are at or just below parallel to the floor. Inhale as you perform this portion of the movement.

3) Ascend until your arms are fully extended overhead and your legs straight by pushing through your heels and pressing the ammo can up towards the ceiling. Allow your neck to hyperextend so that the ammo can has room to clear your chin.

4) Lower the ammo can slowly and transition back into the squatting movement to repeat.

Ammo Can Front Squats

Starting Position: Standing with feet just wider than shoulder width, grasping a 30 lb ammo can (or dumbbell) to your chest.

1) Squat until the crease of your hip is just below your knee cap.

2) Drive back into a standing position with your knees and hips locked out and glutes tight.

Air Squats

Starting Position: Standing with feet more than shoulder width apart.

1) Squat as deep as possible with hips back and heels on the ground. At the same time, lift both arms to shoulder height.

2) Explode from squat position, contracting glutes and return to starting position.

Do 20 reps.

Burpees

Starting Position: Standing with feet shoulder width apart.

1) Perform a deep squat with hands out for balance and smooth transition for next sequence.

2) Place hands flat on the ground and explode back with legs into a push-up position.

3) Perform a push-up and return to the starting position with elbows extended and toes down.

4) Fire your legs in toward your hands, keeping them on the ground and remaining in the squat position.

5) From the squat position explode up and jump with your hands above your head and legs spread eagle in mid-air.

Skaters

Starting Position: Standing with feet shoulder width apart.

1) Shift all your weight onto your right leg, lifting your left foot off the ground. Stabilize on right leg.

2) After a pause, explode off right leg to the left, landing and stabilizing solely on the left while right leg is off the ground. Pause.

3) Explode off left leg, landing and stabilizing on the right while the left is off the ground. Pause and repeat.

Dumbbell Get-Ups

Starting Position: Lying on back holding a 30 lb dumbbell straight above right shoulder with arm straight and vertical. Hold left arm out to the side.

1) Gripping the dumbbell tightly with your right hand, drive your shoulder toward your left hip, shifting onto your left elbow.

2) Shift your weight to your left hand and your right leg.

3) Pull your left leg under your hip into a kneeling position.

4) Shift your weight off of your left hand until your torso is upright. The dumbbell should be balanced over the right shoulder and the right arm.

5) Shift into a lunge position, then stand up.

6) Reverse the steps to lie back down.

7) Switch the dumbbell to the left side and repeat the steps on the opposite side.

Sandbag Get-Ups

Starting Position: Lying on back, with a sandbag on one shoulder.

1) Get up any way you can.

2) One suggestion is to sit up by moving the loaded shoulder toward the opposite leg.

3) Lean on the elbow of your unloaded arm; keeping your core engaged, shift your weight to your hand.

4) Shift your unloaded hip and knee underneath your body into a kneeling position. Your loaded side leg should now be in a lunge position.

5) Stand up from the lunge position and bring your feet underneath your hips. Reverse the process to get back down.

6) Switch the sandbag to the opposite shoulder, and repeat the steps on the opposite side.

Tire Flips

Starting Position: With the tire lying on its side, stand close (as in a deadlift) with your arms inside your legs, and your legs slightly wider than a normal deadlift position and grab the underside of the tire.

1) Grip the tire solidly; lean in, so your chest is above the tire, and your weight on the balls of your feet.

2) Using your legs, begin to stand up while holding the tire firmly.

3) Using a pulling motion, pull the tire upright.

4) Reverse your grip to a pushing position similar to a push-up or bench press, and push the tire away from you explosively. Use your legs and hips to give power to this part of the lift, along with the pushing motion from your arms and chest.

5) Move quickly to the tire, reset your position and continue.

Calisthenics

20-Meter High Crawl

Starting Position: Hands and knees on ground, with feet flexed and your spine forming a straight line from your head to your tailbone.

1) Reach forward with your left hand and right knee. Spine should remain straight.

2) Now move forward with right hand and left knee.

3) Continue for 20 m.

Pull-Ups

Starting Position: A dead hang with hands greater than shoulder width apart, body relaxed.

1) Use both arms and pull your chest up to the bar, pulling your shoulder blades down and back. Keep core and glutes stiff.

2) Actively return to the starting position, pushing chest away from the bar and extending elbows.

Every Minute on the Minute (EMOM) Variation: With a continuously running clock, do one pull-up the first minute, two pull-ups the second minute, three pull-ups the third minute. Continue in this fashion until you are unable to complete the amount of pull-ups required in a one-minute timeframe. Record the number of minutes completed.

Chin-Ups Variation: reverse grip so palms face toward you instead of away from you.

Farmer's Carry

Starting Position: With 50-pound dumbbells (or kettle bells, sandbags, ammo cans, etc.) on the ground just outside hip width, stand with good posture and shoulders back, chest out and feet shoulder width apart, similar to standing at attention.

1) Keeping your back straight and your shoulders set, bend at the knees and sit back until you can grip the objects.

2) While gripping the objects, stand up, keeping your back straight. Your legs will be doing all of the work.

3) Step forward and walk or run at a quick pace while holding the loads and avoid swaying the loads.

Partner Drags

Starting Position: Partner A is seated on ground with forearms clasped over each other and in front of body. Legs are extended. Partner B gets in a dead lift position behind Partner A, snaking his arms under Partner A's arms and grasping Partner A's forearms.

1) Partner A remains relaxed, keeping forearms clasped. Partner B stands up, maintaining a stiff core.

2) Partner B walks backward, maintaining straight torso and firm grasp of Partner A's forearms. Drag for 20 m; then the Partners switch positions.

Walking Ammo Can Lunge

Starting Position: Holding a 30 lb ammo can (or dumbbell) in each hand.

1) Step forward with one leg, and while keeping heel on front leg down and shin relatively vertical, lower yourself until your left thigh is parallel to the floor and your rear knee nearly touches the floor.

2) Stand back up by pushing through the heel of your front leg.

3) Repeat with opposite leg.

Broad Jumps

Starting Position: Standing with heels under hips.

1) Rapidly squat down several inches with hands in front.

2) Explode from the squatting position, and jump as far forward as possible.

3) Land softly with your feet under your hips.

4) Reset immediately and jump again.

Hand Release Push-ups

Starting Position: Front-leaning rest.

1) Drop chest to the ground.

2) Tighten your upper back and pull your hands off of the ground.

3) Push back into the ground and into the push-up position.

Knees to Elbows

Starting Position: In a dead hang from a pull-up bar with hands greater than shoulder width apart.

1) Keeping your upper back tight and your shoulders in the socket, roll your knees up to your elbows in one continuous motion.

2) Drop your feet straight down to avoid kipping. Moving in a smooth and controlled motion will help to avoid swinging from the bar.

USMC Crunches

Starting Position: Lying on your back, with shoulder blades on the deck, knees up and arms across your abdomen.

1) Utilizing a partner or some type of weight to hold your feet on the deck, vigorously drive your upper body off of the ground using your abdominals.

2) Return your shoulders to the deck to complete a crunch

Walking Lunges

Starting Position: Standing with feet hip width apart.

1) Step forward with one leg, keeping your knee tracking over your foot and slightly forward.

2) Drive back into the standing position keeping your body moving in a straight line.

3) Step forward with opposite leg and continue.

Mountain Climbers

Starting Position: Top of push-up position with elbows fully extended.

1) Fire your left leg forward, flexing at the knee while keeping your right leg back, fully extended with knee straight.

2) Quickly alternate, firing right leg forward, flexing at the knee and firing left leg back, fully extended with knee straight.

3) Repeat, alternating legs.

Swimming

Surface swimming

The 10-week fitness program has two surface swimming events per week.

Thursdays involve swimming hard, and working on technique and speed. These workouts are meant to be intense, and they are scheduled after the Wednesday rest day. Each Thursday's swim workout is followed by pull-up work.

Fridays involve longer but less intense swims. These are designed as active recovery and will be completed before or immediately after the Friday interval running.

Submerged swimming

Underwater swim workouts build confidence in the water, and teach you to streamline your body during swim strokes, and gain efficiency. Underwater swimming requires you to develop your breath-holding technique, and to learn to relax and burn less oxygen while swimming.

Preparation and Strokes

Candidates will be expected to perform the USMC side stroke and/or breast stroke as part of the swim assessment.

Side Stroke

The side stroke is conducted as a free-breathing movement, maintaining your face above the surface while your body remains streamlined in the water. You will begin with your chest facing the side of the pool and your arms bent close to your chest. Your head will be turned toward your top shoulder to keep your face out of the water. Your legs will be prepared for a scissor kick movement, with the top leg bent in front of you at 90 degrees like you're sitting in a chair. The bottom leg will extend straight away from your hip, with the knee bent at 90 degrees so that your foot is behind you.

As you begin the movement of the stroke, pull your top hand through the water toward your top thigh, push your bottom hand straight out in front of you in the direction of motion with a flat palm facing the bottom of the pool. Simultaneously, execute the scissor kick where you kick both legs together to initiate your thrust through the water.

Once your legs are together and straight they will align on top of each other and your feet should overlap. Your bottom arm will be fully extended in the intended direction of movement, while your top arm is placed on your upper thigh, streamlined with your body. This is your glide position.

Throughout the entire stroke your face will be oriented perpendicular to the surface of the water so that you can continue to breathe naturally. When you start to lose momentum from your kick, pull your bottom arm in toward your armpit and your top arm up toward your chest. Reset your legs for the scissor kick. This will return you to your starting position, and you can execute the entire stroke again.

CALISTHENICS

Breast Stroke

The breast stroke is not conducted as a free breathing movement. Once you are in a full glide, your face will be in the water at a 45 degree angle to the surface, looking in the intended direction of movement.

You will begin with your elbows bent and your hands near your chest. Take a breath, and as your head goes underwater, extend your arms in front of your body with your hands overlapped and execute the frog kick to propel yourself forward. In order to do the frog kick, pull your heels up to your glutes, kick your legs out at a 45 degree angle to your body, and then bring them together to propel yourself forward. Your legs will be touching and in line behind your body. This is your glide position, with your body parallel to the bottom of the pool, your arms out in front of you, hands overlapping, and legs straight behind you.

Glide for as long as you can before you start to sink, then when you need a breath, draw your arms into your chest to bring your face out of the water so you can breathe. At the same time, draw your heels back up to your gluteus to prepare for another frog kick. As you submerge your head, extend your arms in front of you with hands overlapped, and execute the frog kick to return to your glide position.

Preparing for the Swimming Assessment

You are highly encouraged to:

• Conduct a PFT or a very high intensity cardio workout along with an upper-body workout within 30 to 40 minutes prior to executing your own swim assessment as part of the 10-week fitness program.
• Conduct 15 minutes of treading water: 11 minutes while wearing cammies, with the remaining four minutes making a flotation device with either the top or bottom of the cammies. Should candidates be denied the ability to wear cammies during free swims in military pools, they are highly encouraged to tread water while holding a five-pound brick above their head with one or both hands. This will enhance their ability to use only their legs while treading water, which is necessary for being able to execute the flotation techniques with cammies.
• Wear cammies (no boots) when executing all aquatic training. At times, candidates will be required to tread water for up to 30 minutes while wearing cammies. Should candidates be denied the use of cammies in a military pool during free swim, candidates are highly encouraged to wear a t-shirt.
• Attempt training swims of 500 m, with a target completion time of 15 minutes. This will aid in reaching the goal of swimming 300 m in under 13 minutes.

Treading Water

The initial swim test requires 11 minutes of treading water in your utility uniform, followed by four minutes of survival floating. At times, candidates will be required to tread water for up to 30 minutes while wearing cammies. You have ample time during the 10-week program to learn how to tread water effectively. When performed properly, treading water requires minimal effort, and is vital for life saving and tactical swimming. Treading water with minimal energy demonstrates confidence working in an aquatic environment.

Floating Techniques

There are five different flotation techniques used at A&S: three with the trousers and two with the blouse.

Three Trouser Techniques: Sling, Splash, Blow

SLING Go underwater and take your trousers off. Tie the ankles of your trouser legs together. Bring the top, or waistband, of trousers behind your head and then sling them over your head as hard as possible to trap air inside. Keep the top of the trousers underwater to keep the air in the legs. Pull your head through the legs, with the knot at the ankles behind your head, and lay back and float.

SPLASH Tie trouser legs together at the ankles. Hold opening of trousers at surface of water, and splash water and air into your trousers until the legs are fully inflated. Then put your head through the legs and lay back and float, while keeping the waistband underwater to hold the air in.

BLOW This is the least preferred method because it requires you to take air from your lungs, go underwater, exhale into your trousers, and come back up for air. Hold the opening of the trousers underwater, take a big breath, go underneath the opening of your trousers, exhale all your air into the legs, and repeat until the legs are inflated. Put your head through the opening and lay back and float.

Two Blouse Techniques

1) The first technique is to take a breath, go underwater and exhale the air from your lungs into your blouse on the left side. Come up, get a breath, and exhale into your blouse on the right side. Pull the collar of the blouse tight. The shoulders of the blouse will be inflated. Lean back, cross your legs, and stop moving; you will be able to float with your head above the surface.

2) The second technique is to button the top button of your blouse and slip it over your mouth. Continuously inhale through your nose and exhale through your mouth into your blouse. This will keep a constant flow of air into the shoulders of your blouse and keep you above the surface of the water.

Rucking

Each week has two rucking events, shorter events on Tuesdays and longer rucks on Saturdays.

Tuesdays include 2-4 miles of rucking—1-2 miles before the workout and 1-2 miles after. We suggest 2-4 miles because you may not be ready to move quickly for 4 miles in the beginning. You should move at the fastest pace you can for at least 2 of those miles. Example: Complete the first mile at a warm-up pace and get used to the load, then complete the second mile at your fastest one-mile pace.

Rest, complete the workout, rest again, and then complete the third and fourth miles. This might mean completing the third mile at warm-up pace and the fourth mile at your fastest one-mile pace, or sometimes this is slower depending on the training day. The training is designed to mimic a real world mission. On a mission, you might infiltrate the objective (ruck 1-2 miles), conduct the mission (the workout), and then exfiltrate (ruck 1-2 miles). After Week 3 you should be moving swiftly for 2 miles prior to and after the workout. This should continue until Week 10.

Saturdays include longer-distance rucking, starting with 4 miles in Weeks 1 and 2 and increasing in distance until you reach 12 miles in Week 10. The goal is to improve your endurance while maintaining speed. Your slowest pace should be a 15-minute mile, but you need to average a 13-minute mile or better to be competitive for selection. Candidates use many different techniques to complete these rucks in competitive time.

Rucking Preparation and Guidance

Rucking is simply moving with a ruck of a particular weight for a specified distance. At A&S, candidates are required to move at a minimum pace of 15 minutes per mile. Candidates' rucks will weigh 45 pounds dry. Dry means without water or rations. These rucks will generally be conducted during Phase I on a combination of hardball roads, gravel roads, or trails, all of which are aboard MCB Camp Lejeune, NC.

The ruck candidates use will be issued by MARSOC Central Issuing Facility (CIF) during check-in at A&S. The instructor cadre will explain the how the gear is configured, and the standard operating procedures (SOPs) for ruck-based training prior to any event. It is the candidate's responsibility to ensure that his ruck meets the minimum weight requirements (45 pounds) and contains the specified gear needed for the event that is briefed, or based on the SOP.

All candidates will carry rubber training M-16 rifles weighing approximately 10 pounds. Candidates may not use a sling during any of the selection phases.

Gear

To complete the 10-week program, you will need a ruck (either issued or purchased). We recommend you purchase the type used at A&S (external frame Mountain Ruck, or large ALICE Pack) so you will be prepared for the conditions at A&S.

Footwear

During selection you will use Marine Corps-approved combat boots. The type of socks you wear are your choice, as long as they meet regulations for color. We advise you to train in the same boots and socks you will wear at A&S.

Uniform

Candidates will wear the seasonally appropriate, or specified USMC digital utility uniforms during A&S. When training in locations outside of military bases, this may not be functional or appropriate. If you are not able to wear USMC digital utilities uniform where you train, purchase military-style fatigues that fit similarly to your standard uniform. You want clothing and gear that mimic what you will wear at A&S.

Rubber Rifles

Ideally, you will be able to obtain and use a rubber rifle. NOTE: We advise you to only carry one on a military installation. It may not be feasible or smart to ruck with one outside of military installations. Off base you can use several other objects that are close to a rifle's shape and weight to mimic carrying a training rifle. Be creative but considerate to the public.

1) Exercise weight poles of the same length and weight as a rifle.
2) A 10-pound sledge hammer or similar implement.
3) Several pieces of rebar taped or tied together in the length and weight needed.

Rucking for Distance

The minimum per mile time for rucking at A&S is 15 minutes per mile. This is with the 45-pound ruck (plus water and chow carried in the ruck), and a rubber rifle with no sling.

You will cover 8-12 miles on the rucks at A&S.

- 8 miles: maximum time 2 hours
- 10 miles: maximum time 2 hours 30 minutes
- 12 miles: maximum time 3 hours

Minimum times are indicated so that you understand what the bare minimum standards are. The times above are not meant to be training goals. Your training goal is to move much faster than the minimum times. The fastest recorded 12-mile ruck time at A&S is 1 hour 47 minutes, which is just under a 9-minute mile pace. That's 6 minutes per mile below the minimum 15-minute mile pace.

NOTE: All physical training events at A&S are competitive. Your goal is to be the best candidate you can be. Your ruck times are part of you overall score. Move as quickly as you can and do your best to beat the minimum times.

Candidates use many different techniques in order to move fast enough to be competitive. Here are a few strategies.

CALISTHENICS

Rucking Strategies

Mile on/ mile off

Move at a certain pace for one mile then a slower pace for the next mile to recover. Example: one mile at a 12-minute mile pace, then one mile at a 13-minute mile pace.

Run the terrain

Run the down hills and straightaways, but walk swiftly up inclines and hills.

Time or distance ratios

Run for three minutes, walk for one minute. Distance ratios made up on the fly: run to the next bend in the road; walk till to the next bend in the road.

NOTE: To be competitive as a candidate it is vital for you to record your times on all events for rucking, swimming and running. It is also wise to record your times for the workouts if only for planning purposes. This trains you to move with a purpose and get the work done. It will carry over into your attitude and ability at A&S. Your numbers and times are important for you to understand where you are in training, and to know if you are getting better or if your progress is stalled. Awareness of your capabilities is crucial for success as a candidate.

Weeks 6, 8, and 10 are meant as training checks for your progress. The training schedule for those weeks is built to mimic the order of events at A&S Phase I. Fridays will be a Marine Corps PFT followed by the swim test. Saturday will be a long ruck. Prepare accordingly.

NOTE: If you are having trouble making these times on either the Tuesday or Saturday ruck events, you should consider skipping a day of training for added recovery. Ensure you are eating enough, drinking sufficient water, and getting enough sleep.

Running

This 10-week program includes distance runs and sprints.

Mondays include three miles of running, either three miles before your workout or 1.5 miles before and after the workout. These runs should be completed within a minute of your PFT 3-mile time.

Fridays include 200-, 400-, and 800-meter sprints. While 800 m is technically not a sprint, the key is to run faster than PFT pace.

Auxiliary running: Some workouts include jogs and sprints. This is meant to train your body to handle multiple types of efforts in a single workout. Although the running is not the main focus of the workout, it will help you on your PFT time as well as assist with rucking.

Calisthenics

MOVEMENT CARD ❶ CORE WORK

Movement (page #)	Distance/reps/time	
1 **MARSOC In Test Preparation** - Do Three Rounds		
Pull-Ups (18)	8-12 reps (no kipping)	☐
Crunches (22)	20-40 reps	☐
Burpees (15)	10 reps	☐
2 **Rest**	2 minutes	
3 **Core Work** - Do Five Rounds		
Front Plank (12)	30-60 seconds	☐
Side Plank (12)	30-60 seconds	☐
Knees to Elbow* (21)	10-15 reps	☐
Dumbbell Get-Up (16)	10 reps	☐
Flutter Kicks (12)	20 reps (4 count)	☐
4 **Rest**	2 minutes	
5 **Crunches*** (22)	2 x 100 (4 count)**	☐

* If pull-up bars are not available, conduct regular sit-ups.

** Perform the 100 reps in as few sets as possible. Rest 2-3 minutes between sets.

MOVEMENT CARD ❷ RUCK BASED

Note: do not use a ruck unless indicated

Movement (page #)	Distance/reps/time	
1 **Dynamic Warm Up**		
Lunges With Counter Rotation (13)	10 reps	☐
Skaters (16)	30 reps	☐
Broad Jumps (20)	10 reps	☐
Mountain Climbers (22)	20 (4 count)	☐
Skaters (16)	30 reps	☐
Dumbbell Get-Ups (16)	10 reps	☐
2 **Ruck** (26)	1-2 miles	☐
3 **Planks / Frog Squats** - Do Two Rounds		
Forward Plank (12)	Hold for 1 minute	☐
Side Plank (12)	Hold for 30 seconds each side	☐
Frog Squats (13)	1 minute	☐
4 **Rest**	2 minutes	

continued...

MOVEMENT CARD ❷ RUCK BASED continued

Movement (page #)	Distance/reps/time	
5 **MARSOC In Test Preparation** - Do Five Rounds at a Brisk Pace With Good Form. Rest as Necessary Between Sets.		
Crunches (22)	30 reps	☐
Pull-Ups (18)	5-10 reps	☐
Jog (29)	100-200 m	☐
6 **Rest**	2 minutes	
7 **Workout** - Do Four Rounds		
High Crawl (18)	20 m	☐
Partner Drag (19)	20 m	☐
Walking Ammo Can Lunge (20)	20 m	☐
Tire Flips (17)	5-10, with Ruck	☐
Ammo Can Overhead Lunge (14)	25m, with Ammo Can	☐
8 **Ruck** (26)	1-2 miles	☐

MOVEMENT CARD ❸ MARSOC IN TEST PREP AND CORE

Movement (page #)	Distance/reps/time	
1 **Planks and Frog Squats**		
Forward Plank (12)	Hold for 1 minute, 30 seconds	☐
Side Plank (12)	Hold for 1 minute each side	☐
Frog Squats (13)	1 minute	☐
2 **MARSOC In Test Preparation** - Do Four Rounds		
Hand Release Push-Ups (21)	10-15 reps	☐
Pull-Ups (18)	8-10 reps (no kipping)	☐
Crunches (22)	20-30 reps	☐
3 **Rest**	2 minutes	
4 **Crunches** (22)	Max reps for 2 minutes, 1 minute, 30 seconds, with 2-3 minutes rest between sets	☐

MOVEMENT CARD ❹ RUCK BASED STENGTH

Movement (page #)	Distance/reps/time	
1 **MARSOC In Test Preparation** - Do Three Rounds		
Pull-Ups (18)	10-12 reps	☐
Crunches (22)	15-20 reps	☐
Burpees (15)	10	☐
2 **Rest**	2 minutes	
3 **Workout** - Do Four Rounds		
Ammo Can Front Squats (15)	10 reps	☐
Walking Lunges (22)	20-30 reps	☐
Ammo Can Thrusters (14)	10 reps	☐
Dumbbell Get-Ups (16)	5-10	☐

Calisthenics

MOVEMENT CARD ❺ TOTAL BODY

Movement (page #)	Distance/reps/time	
1 Planks and Frog Squats		
Forward Plank (12)	Hold for 1 min	☐
Side Plank (12)	Hold for 30 sec each side	☐
Frog Squats (12)	30 sec	☐
2 Warm Up - Do Three Rounds		
Pull-Ups (18)	3 reps (no kipping)	☐
Push-Ups (13)	10 reps	☐
Air Squats (15)	10 reps	☐
Skaters (16)	10 reps	☐
3 Workout - Do Ten-to-Twenty Rounds		
Pull-Ups (18)	5 reps (no kipping)	☐
Push-Ups (13)	10 reps	☐
Air Squats (15)	15 reps	☐
4 Rest	2 minutes	
5 Farmer's Carry (15)	5-10 reps of 50 m, with 50 pounds or more in each hand. Rest 30 seconds between reps.	

Calisthenics

MOVEMENT CARD ❻ RUCK CORE

Movement (page #)	Distance/reps/time	
1 **Workout** - Do Five Rounds		
Forward Plank (12)	Hold for 1 min	☐
Air Squat (15)	10 reps	☐
Dumbbell Get-Ups (16)	10 reps	☐
Burpees (15)	10 reps	☐
2 **Rest**	2 minutes	
3 **In Test Preparation** - rest 2-3 min between sets		
Crunches / Forward Plank	2 min max reps / 1 min	☐
Crunches / Forward Plank	1 min max reps / 1 min	☐
Crunches / Forward Plank	30 sec max reps / 1 min	☐
4 **Pyramid Pull-Ups** (18)	1, 2, 3, 4, 5, 4, 3, 2, 1 (25 total)	☐

SWIM CARD ❶

Movement (page #)	Distance/reps/time	
1 **Warm-Up** - Do Four Rounds		
Swim (23)	25 m	☐
Push-Ups (13)	10 reps	☐
Air Squats (15)	10 reps	☐
Crunches (22)	10 reps	☐

Perform the following in cammies

2 **Swim** (23)	5 x 100 m with 30 second break between each round	☐
3 **Tread Water** (25)	3 x 30 seconds (15 seconds with left hand held above water, 15 seconds with right hand held above water), with 30 seconds between each round	☐
4 **Workout** - Do Four Rounds for Time		
Swim (23)	50 m	☐
Push-Ups (13)	10	☐
Flutter Kicks (12)	15	☐
5 **Pull-Ups** (18)	3 sets at 75% max reps	☐

Calisthenics

SWIM CARD ❷

Movement (page #)	Distance/reps/time	
1 **Warm-Up** - Do Four Rounds		
Swim (23)	25 m	☐
Push-Ups (13)	10 reps	☐
Air Squats (15)	10 reps	☐
Crunches (22)	10 reps	☐

Perfom the following in cammies

2 **Swim** (23)	300 m for time	☐
3 **Tread Water** (25)	10 minutes	☐
4 **Workout** - Do Four Rounds for Time		
Swim (23)	50 m	☐
Push-Ups (13)	10	☐
Flutter Kicks (12)	15	☐
5 **EMOM Pull-Ups** (18)		☐

SWIM CARD ❸

Movement (page #)	Distance/reps/time	
1 **Warm-Up** - Do Three Rounds		
Air Squats (15)	10 reps	☐
Push-Ups (13)	8 reps	☐
Crunches (22)	8 reps	☐
Flutter Kicks (12)	10 reps	☐

Perfom the following in cammies

2 **Swim** (23)	300 m for time	☐
3 **Tread Water** (25)	15 Minutes	☐
4 **Workout** - Do Five Rounds for Time		
Swim (23)	200 m	☐
Air Squats (15)	25	☐
5 **Pull-Ups Pyramid** (18)	1,2,3,4,5,4,3,2,1	☐
6 **Chin-Ups Pyramid** (18)	1,2,3,4,5,4,3,2,1	☐
7 **Pull-Ups Pyramid** (18)	1,2,3,4,5,4,3,2,1	☐

SWIM CARD ❹

Movement (page #)	Distance/reps/time	
1 **Warm-Up** - Do Four Rounds		
Swim (23)	50 m	☐
Push-Ups (13)	10 reps	☐
Air Squats (15)	10 reps	☐
Crunches (22)	10 reps	☐
Perfom the following in Cammies		
2 **Swim** (23)	300 m	☐
3 **Tread Water** (25)	4 x 2 minutes (1 minute with left hand held above water, 1 minute with right hand held above water), with 30 seconds between each round	☐
4 **Tread Water** (25)	10 minutes	☐
5 Do Four Rounds for Time		
Swim (23)	100 m	☐
Push-Ups (13)	15	☐
Flutter Kicks (12)	20	☐
6 **Pull-Ups** (18)	4 x 10 reps (no kipping)	☐

Calisthenics

SWIM CARD ❺

Movement (page #)	Distance/reps/time	
1 **Warm-Up** - Do Four Rounds		
Swim (23)	25 m	☐
Push-Ups (13)	10 reps	☐
Air Squats (15)	10 reps	☐
Crunches (22)	10 reps	☐

Perfom the following in Cammies

2 **Swim** (23)	500 m for time	☐
3 **Tread Water** (25)	30 minutes	☐
4 Do Three Rounds, with 1 Minute Rest Between Rounds		
Underwater Swim (23)	length of pool underwater	☐
Surface Swim (23)	25 m	☐
6 **Pull-Ups** (18)	3 sets of 75% max reps	☐

Calf

Starting Position: Lying on your back with the resistance band wrapped around your right forefoot.

1) Lift right leg to a 30 degree angle and flex foot while you pull the band back. At end of range of motion, exhale and gently pull the band until you feel a stretch. Hold for 2 seconds.

2) Inhale and now point your toes away from your shin. Pause and repeat.

Do 10 reps then repeat on opposite side.

Hamstring

Starting Position: Lying on your back with the resistance band wrapped around your right forefoot.

1) Lift your right leg, keeping knee straight as you pull the band. At end of range of motion, exhale and gently pull the band until you feel a stretch. Hold 2 seconds.

2) Inhale and return to starting position.

Do 10 reps then repeat on opposite side.

IT Band

Starting Position: Lying on your back with the resistance band wrapped around your foot, hold the band in the hand opposite the leg you are stretching. Other hand is flat on ground.

1) While keeping shoulders on the ground and non-working leg stationary with toes pointed to sky, fire inner thigh muscles to bring working leg across your body as you pull with the band. At the end of the range of motion, give a gentle stretch for 2 seconds.

2) Relax and return leg to starting position.

Do 10 reps then repeat on opposite side.

Groin

Starting Position: Lying on your back with the resistance band wrapped around your foot, hold the band with the same side hand as working leg.

1) With non-working leg stationary with toes pointed skyward, sweep working leg away from body, keeping knee straight. At end of range of motion, give band gentle pull to assist in stretch. Hold for two seconds while exhaling. Pause, inhale and return to starting position.

Do 10 reps then repeat on opposite side.

Quadriceps

Starting Position: Lying on your stomach with band wrapped around forefoot of working leg and band pulled over same side shoulder.

1) Stationary leg remains straight while firing the working leg hamstring.

2) At end of range of motion, fire glute and lift working leg off ground. After lifting the leg, gently pull the band to provide a gentle stretch, holding for 2 seconds while you exhale. Pause, relax and return to starting position.

Do 10 reps then repeat on opposite side.

Triceps

Starting Position: Reach over your shoulder with your right hand, elbow pointing up to take the band. With left hand behind your back, grab band, palm facing out.

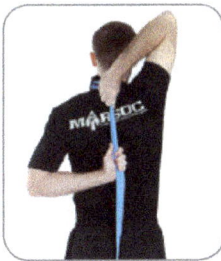

1) Reach with right hand down the spine. At end of range of motion, gently pull band with left hand for a 2-second stretch. After 2 seconds inhale and return hands to starting position.

Exchange sides and repeat.

Rotator Cuff

Starting Position: Lying on your right side with your arm in a 90/90 position.

1) Externally rotate your right shoulder to try to put the back of your right hand on the ground, maintaining a 90-degree flex on your elbow.

2) Now internally rotate your right shoulder and attempt to put your right palm on the ground in front of your belly button. Hand placement should be behind watch. At end of range of motion, gently assist stretch with your left hand. Exhale and hold for 2 seconds. Relax, inhale and return to starting position.

Do 10 reps then repeat on opposite side.

Quadruped Thoracic Spine

Starting Position: Start on hands and knees, with back straight. Sit back with butt on heels.

1) In a continuous swinging motion reach across the body and under the left arm with the right hand with palm up and then pull the hand back across to the back of the head flexing to the right.

Do 5 reps then repeat on opposite side.

Middle Back

Starting Position: Lying on your left side with bottom left leg extended, right arm extended and reaching out, right knee flexed up toward chest being held down by the bottom left hand.

1) Open your shoulders by rotating torso to the right, attempting to put upper back and right shoulder and right arm on the ground as you exhale. Hold for 2 seconds then return to starting position while inhaling. Do 10 reps then repeat on opposite side.

Calf

Starting Position: Place foam roll under right calf. Place body weight on right leg.

1) Roll calf by using arms to let lower leg glide up and down the roll. Perform 20-30 rolls and perform 20-30 slow rolls over any tender areas.

2) Switch legs and repeat.

Hamstring

Starting Position: Place foam under right hamstring. Place all body weight on roll.

1) Roll hamstring by using arms to let your body glide up and down roll. Perform 20-30 slow rolls on any trigger spots.

2) Switch legs and repeat.

IT Band

Starting Position: With foam roll underneath you, lean on your right side, supported by your forearm.

1) Do 20-30 rolls for each 1/3 of the leg - hip bone to 1/3 down leg, middle 1/3 of leg, knee to ankle. Perform 20-30 slow rolls on any trigger spots.

2) Switch legs and repeat.

Quadriceps

Starting Position: Lying face down with roller under quads, and arms extended.

1) Lower arms and lift right leg off the ground and perform 20-30 rolls. Perform 20-30 slow rolls on any trigger spots.

2) Relax, switch legs and repeat.

Groin

Starting Position: Lying down with right leg straight and left leg at 45-degree angle supported on your arms, slightly bent. Foam roller positioned along groin and left inner thigh.

1) Roll for 20-30 repetitions along groin and inner thigh. Perform 20-30 slow rolls on any trigger spots.

2) Switch legs and repeat.

Glute

Starting Position: Sitting with right leg extended, left leg bent, with foam roller under right glute. Hands placed behind you.

1) Roll for 20-30 reps along groin and inner thigh. Perform 20-30 slow rolls on any trigger spots.

2) Switch legs and repeat.

Middle and Upper Back

Starting Position: Lying on back with foam roller at base of neck.

1) Roll from the base of your neck to the middle of your back. You can support your head with your hands if you prefer.

2) Perform 20-30 rolls, relax and repeat.

Lats

Starting Position: Lying on your side with the foam roller just below your arm.

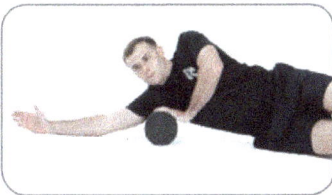

1) Do 20-30 rolls from just under your armpit and along your rib cage and perform 20-30 slow rolls on any trigger spots.

2) Switch sides and repeat.

Hip External Rotator

Starting Position: Sitting on foam roller, cross right ankle onto left knee and rotate your body so that all weight is on right glute. Stabilize yourself with your right hand.

1) Roll back and forth on right glute for 20-30 rolls and perform 20-30 slow rolls on any trigger spots.

2) Switch legs and repeat.

Hip Flexor

Starting Position: Lying facedown on forearms with foam roller just below right hip flexor.

1) Using arms, glide hip flexor over roll for 20-30 reps, and perform 20-30 slow rolls on any trigger spots.

2) Shift weight to left hip flexor and repeat.

Hip Adductor

Starting Position: Face down with leg raised toward shoulder and roller under left leg.

1) Perform 20-30 rolls on inside of thigh from knee to pelvis. Perform 20-30 slow rolls over any tender areas.

2) Switch legs and repeat.

Post-Workout CARD

AIS: Utilize band for all exercises except Rotator Cuff, Quadruped Thoracic Spine, and AIS Middle Back stretch; all are 10 reps each, start with left side and do 1-4 then do right side 1-4. Exercises 6-9 do both sides L/R then go to the next exercise.

1	**Calf** (42)	Pull, Relax, 1; Pull, Relax, 2;...10
2	**Hamstring** (42)	Pull, Relax, 1; Pull, Relax, 2;...10
3	**IT Band** (42)	Pull, Relax, 1; Pull, Relax, 2;...10
4	**Groin** (43)	Pull, Relax, 1; Pull, Relax, 2;...10
5	**Quadriceps** (43)	Up, Pull, Relax 1; Up, Pull, Relax, 2;...10
6	**Rotator Cuff** (44)	Down, Back, 1; Down, Back, 2;...10
7	**Quadruped Thoracic Spine** (44)	Count on your own
8	**Middle Back** (44)	Up, Back, 1; Up, Back, 2;...10
9	**Triceps** (43)	Down, Relax, 1; Down, Relax, 2;...10

ROLLER: All are 20-30 rolls each

1	**Calf** (45)	7	**Middle and Upper Back** (47)
2	**Hamstring** (45)	8	**Lats** (47)
3	**IT Band** (45)	9	**Hip External Rotator** (47)
4	**Quadriceps** (46)	10	**Hip Flexor** (48)
5	**Groin** (46)	11	**Hip Adductor** (48)
6	**Glute** (46)		

Nutrition

The 10-week program is intense. It requires that you eat properly in order to perform well. Height and weight differences affect caloric intake, which can be as high as 5,000 calories a day or as low as 3,000. Ensure you eat enough food to perform well on the events, as well as to recover from them. Your body also needs macronutrients to function well. You need adequate protein to repair and build muscle, fat to provide energy and support molecular functions, and carbohydrates to provide energy stores for training, recovery and recuperation. Look for the healthiest types of macronutrients possible. Although most athletes try to eat healthy and stay away from processed foods, your caloric demands may require you to eat almost anything that is available to you.

Hydration

To perform well during the 10-week program, you have to take in enough fluid to allow for proper bodily function, including removal of wastes and movement of nutrients to tissues. Ensure you drink enough fluid to help digest all of the food you are eating and to compensate for what you will sweat out, which may vary depending on whether you are in hot or cold environments. Avoid drinking too much water, as this can lead to a serious condition called hypernatremia, during which minerals, electrolytes and salts needed for normal functions are flushed out of your body. Try to eat frequently throughout the day as you are hydrating; this keeps the needed minerals, salts and electrolytes in your body. Check the color of your urine often; it should be clear and plentiful. If you are adequately hydrated, you may wake in the middle of the night to urinate.

Some vitamins and food will color your urine; keep that in mind as you judge your hydration. If you do not have enough water in your body, you may experience muscle cramps, soreness, nausea, and lethargy. If your urine output decreases, you should step up your hydration.

Foot Care

You must take care of your feet. This includes properly clipping your toenails and keeping your feet clean. Change your socks and powder your feet between ruck events on Tuesdays and Saturdays. If you have problems with blisters or other serious issues such as painful, swollen feet, see your medical provider. Do your best to avoid blisters and other irritations by changing socks frequently and breaking in your boots properly. If you feel hot spots or tender spots, change socks, apply foot powder, or use mole skin or similar products.

Recovery

You should include rest, stretching, hydration, and recovery in your training. Put time and effort into mastering the strength and fitness exercises as well as recovery activities, which you should do several times a day on workout days as well as on Wednesday, which are rest days. If you generally feel good after completing the daily workout (and accomplishing the suggested times and repetitions), then active recovery is a good idea. If you are exhausted, inactive recovery or sleep is a better option.

Active recovery increases blood flow to sore muscles. This can be a light jog, light calisthenics, or simply walking or playing a sport. If inactive recovery is required, keep doing your daily workouts, but don't engage in other physical training if at all possible.

Adequate sleep is essential to your performance. Sleep allows your body to rebuild and strengthen tissues. Try to go to sleep fully hydrated, and with a full stomach; this facilitates repair and recovery.

With proper nutrition, hydration and recovery, you will have a greater chance of success at selection.

WEEK ❶	⏱ Time

Sun	**Rest, stretch, hydrate & recover** (Movement Prep and Post-Workout Cards)		
Mon	☐ **Movement Prep Card** (pg 11)	h:	m:
	☐ **1.5-mile run** (pg 29)	h:	m:
	☐ **Movement Card #1** (pg 30)	h:	m:
	☐ **1.5-mile run** (pg 29)	h:	m:
	☐ **Post-Workout Card** (pg 49)	h:	m:
Tue	☐ **Movement Prep Card** (pg 11)	h:	m:
	☐ **1-2 mile ruck, faster than 13 minutes per mile** (pg 26)	h:	m:
	☐ **Movement Card #2** (pg 31)	h:	m:
	☐ **1-2 mile ruck, faster than 13 minutes per mile** (pg 26)	h:	m:
	☐ **Post-Workout Card** (pg 49)	h:	m:
Wed	**Rest, stretch, hydrate & recover** (Movement Prep and Post-Workout Cards)		
Thu	☐ **Movement Prep Card** (pg 11)	h:	m:
	☐ **Swim Card #1** (pg 37)	h:	m:
	☐ **Post-Workout Card** (pg 49)	h:	m:
Fri	☐ **Movement Prep Card** (pg 11)	h:	m:
	☐ **4 x 400 m run** (pg 29)	h:	m:
	☐ **2 x 200 m swim** (pg 23)	h:	m:
	☐ **Post-Workout Card** (pg 49)	h:	m:
Sat	☐ **Movement Prep Card** (pg 11)	h:	m:
	☐ **4-mile ruck, 1 hr or less** (pg 26)	h:	m:
	☐ **Assess feet/gear** (pg 51)	h:	m:
	☐ **Post-Workout Card** (pg 49)	h:	m:

WEEK ❶ Notes (additional exercises, nutrition, sleep, fatigue)

Sun

Mon

| Energy level | 10 | 9 | 8 | 7 | 6 | 5 | 4 | 3 | 2 | 1 |

Tue

| Energy level | 10 | 9 | 8 | 7 | 6 | 5 | 4 | 3 | 2 | 1 |

Wed

| Energy level | 10 | 9 | 8 | 7 | 6 | 5 | 4 | 3 | 2 | 1 |

Thu

| Energy level | 10 | 9 | 8 | 7 | 6 | 5 | 4 | 3 | 2 | 1 |

Fri

| Energy level | 10 | 9 | 8 | 7 | 6 | 5 | 4 | 3 | 2 | 1 |

Sat

| Energy level | 10 | 9 | 8 | 7 | 6 | 5 | 4 | 3 | 2 | 1 |

WEEK ❷	⏱ Time

Sun	Rest, stretch, hydrate & recover (Movement Prep and Post-Workout Cards)		
Mon	☐ **Movement Prep Card** (pg 11)	h:	m:
	☐ **3-mile run** (pg 29)	h:	m:
	☐ **Movement Card #3** (pg 39)	h:	m:
	☐ **Post-Workout Card** (pg 49)	h:	m:
Tue	☐ **Movement Prep Card** (pg 11)	h:	m:
	☐ **1-2 mile ruck, faster than 13 minutes per mile** (pg 26)	h:	m:
	☐ **Movement Card #4** (pg 40)	h:	m:
	☐ **Post-Workout Card** (pg 49)	h:	m:
Wed	Rest, stretch, hydrate & recover (Movement Prep and Post-Workout Cards)		
Thu	☐ **Movement Prep Card** (pg 11)	h:	m:
	☐ **Swim Card #2** (pg 38)	h:	m:
	☐ **Post-Workout Card** (pg 49)	h:	m:
Fri	☐ **Movement Prep Card** (pg 11)	h:	m:
	☐ **9 x 200 m run** (pg 29)	h:	m:
	☐ **500 m swim** (pg 23)	h:	m:
	☐ **Post-Workout Card** (pg 49)	h:	m:
Sat	☐ **Movement Prep Card** (pg 11)	h:	m:
	☐ **4-mile ruck, 1 hr or less** (pg 26)	h:	m:
	☐ **Assess feet/gear** (pg 51)	h:	m:
	☐ **Post-Workout Card** (pg 49)	h:	m:

WEEK ❷ Notes (additional exercises, nutrition, sleep, fatigue)

Sun

Mon

Energy level	10	9	8	7	6	5	4	3	2	1

Tue

Energy level	10	9	8	7	6	5	4	3	2	1

Wed

Energy level	10	9	8	7	6	5	4	3	2	1

Thu

Energy level	10	9	8	7	6	5	4	3	2	1

Fri

Energy level	10	9	8	7	6	5	4	3	2	1

Sat

Energy level	10	9	8	7	6	5	4	3	2	1

WEEK ❸	⏱ Time

Sun	**Rest, stretch, hydrate & recover** (Movement Prep and Post-Workout Cards)	
Mon	☐ **Movement Prep Card** (pg 11)	h: m:
	☐ **1.5-mile run** (pg 29)	h: m:
	☐ **Movement Card #5** (pg 35)	h: m:
	☐ **1.5-mile run** (pg 29)	h: m:
	☐ **Post-Workout Card** (pg 49)	h: m:
Tue	☐ **Movement Prep Card** (pg 11)	h: m:
	☐ **1-2 mile ruck, faster than 13 minutes per mile** (pg 26)	h: m:
	☐ **Movement Card #6** (pg 36)	h: m:
	☐ **1-2 mile ruck, faster than 13 minutes per mile** (pg 26)	h: m:
	☐ **Post-Workout Card** (pg 49)	h: m:
Wed	**Rest, stretch, hydrate & recover** (Movement Prep and Post-Workout Cards)	
Thu	☐ **Movement Prep Card** (pg 11)	h: m:
	☐ **Swim Card #3** (pg 39)	h: m:
	☐ **Post-Workout Card** (pg 49)	h: m:
Fri	☐ **Movement Prep Card** (pg 11)	h: m:
	☐ **5 x 400 m run** (pg 29)	h: m:
	☐ **2 x 400 m swim** (pg 23)	h: m:
	☐ **Post-Workout Card** (pg 49)	h: m:
Sat	☐ **Movement Prep Card** (pg 11)	h: m:
	☐ **5-mile ruck, 1:30 or less** (pg 26)	h: m:
	☐ **Assess feet/gear** (pg 51)	h: m:
	☐ **Post-Workout Card** (pg 49)	h: m:

WEEK ❸ Notes (additional exercises, nutrition, sleep, fatigue)

Sun

Mon

| Energy level | 10 | 9 | 8 | 7 | 6 | 5 | 4 | 3 | 2 | 1 |

Tue

| Energy level | 10 | 9 | 8 | 7 | 6 | 5 | 4 | 3 | 2 | 1 |

Wed

| Energy level | 10 | 9 | 8 | 7 | 6 | 5 | 4 | 3 | 2 | 1 |

Thu

| Energy level | 10 | 9 | 8 | 7 | 6 | 5 | 4 | 3 | 2 | 1 |

Fri

| Energy level | 10 | 9 | 8 | 7 | 6 | 5 | 4 | 3 | 2 | 1 |

Sat

| Energy level | 10 | 9 | 8 | 7 | 6 | 5 | 4 | 3 | 2 | 1 |

WEEK ❹	⏱ Time

		Time
Sun	**Rest, stretch, hydrate & recover** (Movement Prep and Post-Workout Cards)	
Mon	☐ **Movement Prep Card** (pg 11)	h:　　m:
	☐ **3-mile run** (pg 29)	h:　　m:
	☐ **Movement Card #1** (pg 30)	h:　　m:
	☐ **Post-Workout Card** (pg 49)	h:　　m:
Tue	☐ **Movement Prep Card** (pg 11)	h:　　m:
	☐ **1-2 mile ruck, faster than 13 minutes per mile** (pg 26)	h:　　m:
	☐ **Movement Card #2** (pg 31)	h:　　m:
	☐ **1-2 mile ruck, faster than 13 minutes per mile** (pg 26)	h:　　m:
	☐ **Post-Workout Card** (pg 49)	h:　　m:
Wed	**Rest, stretch, hydrate & recover** (Movement Prep and Post-Workout Cards)	
Thu	☐ **Movement Prep Card** (pg 11)	h:　　m:
	☐ **Swim Card #4** (pg 40)	h:　　m:
	☐ **Post-Workout Card** (pg 49)	h:　　m:
Fri	☐ **Movement Prep Card** (pg 11)	h:　　m:
	☐ **10 x 200 m run** (pg 29)	h:　　m:
	☐ **800 m swim** (pg 23)	h:　　m:
	☐ **Post-Workout Card** (pg 49)	h:　　m:
Sat	☐ **Movement Prep Card** (pg 11)	h:　　m:
	☐ **5-mile ruck, 1:25 or less** (pg 26)	h:　　m:
	☐ **Assess feet/gear** (pg 51)	h:　　m:
	☐ **Post-Workout Card** (pg 49)	h:　　m:

WEEK ❹ Notes (additional exercises, nutrition, sleep, fatigue)

Sun

Mon

| Energy level | 10 | 9 | 8 | 7 | 6 | 5 | 4 | 3 | 2 | 1 |

Tue

| Energy level | 10 | 9 | 8 | 7 | 6 | 5 | 4 | 3 | 2 | 1 |

Wed

| Energy level | 10 | 9 | 8 | 7 | 6 | 5 | 4 | 3 | 2 | 1 |

Thu

| Energy level | 10 | 9 | 8 | 7 | 6 | 5 | 4 | 3 | 2 | 1 |

Fri

| Energy level | 10 | 9 | 8 | 7 | 6 | 5 | 4 | 3 | 2 | 1 |

Sat

| Energy level | 10 | 9 | 8 | 7 | 6 | 5 | 4 | 3 | 2 | 1 |

WEEK ❺	⏱ Time
Sun	**Rest, stretch, hydrate & recover** (Movement Prep and Post-Workout Cards)

		h:	m:
Mon	☐ **Movement Prep Card** (pg 11)	h:	m:
	☐ **1.5-mile run** (pg 29)	h:	m:
	☐ **Movement Card #3** (pg 39)	h:	m:
	☐ **1.5-mile run** (pg 29)	h:	m:
	☐ **Post-Workout Card** (pg 49)	h:	m:
Tue	☐ **Movement Prep Card** (pg 11)	h:	m:
	☐ **1-2 mile ruck, faster than 13 minutes per mile** (pg 26)	h:	m:
	☐ **Movement Card #4** (pg 40)	h:	m:
	☐ **1-2 mile ruck, faster than 13 minutes per mile** (pg 26)	h:	m:
	☐ **Post-Workout Card** (pg 49)	h:	m:
Wed	**Rest, stretch, hydrate & recover** (Movement Prep and Post-Workout Cards)		
Thu	☐ **Movement Prep Card** (pg 11)	h:	m:
	☐ **Swim Card #5** (pg 41)	h:	m:
	☐ **Post-Workout Card** (pg 49)	h:	m:
Fri	☐ **Movement Prep Card** (pg 11)	h:	m:
	☐ **3 x 800 m run** (pg 29)	h:	m:
	☐ **2 x 300 m swim** (pg 23)	h:	m:
	☐ **Post-Workout Card** (pg 49)	h:	m:
Sat	☐ **Movement Prep Card** (pg 11)	h:	m:
	☐ **6-mile ruck, 1:40 or less** (pg 26)	h:	m:
	☐ **Assess feet/gear** (pg 51)	h:	m:
	☐ **Post-Workout Card** (pg 49)	h:	m:

WEEK ❺ Notes (additional exercises, nutrition, sleep, fatigue)

Sun

Mon

| Energy level | 10 | 9 | 8 | 7 | 6 | 5 | 4 | 3 | 2 | 1 |

Tue

| Energy level | 10 | 9 | 8 | 7 | 6 | 5 | 4 | 3 | 2 | 1 |

Wed

| Energy level | 10 | 9 | 8 | 7 | 6 | 5 | 4 | 3 | 2 | 1 |

Thu

| Energy level | 10 | 9 | 8 | 7 | 6 | 5 | 4 | 3 | 2 | 1 |

Fri

| Energy level | 10 | 9 | 8 | 7 | 6 | 5 | 4 | 3 | 2 | 1 |

Sat

| Energy level | 10 | 9 | 8 | 7 | 6 | 5 | 4 | 3 | 2 | 1 |

WEEK ❻	⏱ Time

Sun	**Rest, stretch, hydrate & recover** (Movement Prep and Post-Workout Cards)	
Mon	☐ **Movement Prep Card** (pg 11)	h:　m:
	☐ **3-mile run** (pg 29)	h:　m:
	☐ **Movement Card #5** (pg 35)	h:　m:
	☐ **Post-Workout Card** (pg 49)	h:　m:
Tue	☐ **Movement Prep Card** (pg 11)	h:　m:
	☐ **1-2 mile ruck, faster than 13 minutes per mile** (pg 26)	h:　m:
	☐ **Movement Card #6** (pg 36)	h:　m:
	☐ **1-2 mile ruck, faster than 13 minutes per mile** (pg 26)	h:　m:
	☐ **Post-Workout Card** (pg 49)	h:　m:
Wed	**Rest, stretch, hydrate & recover** (Movement Prep and Post-Workout Cards)	
Thu	**Rest, stretch, hydrate & recover** (Movement Prep and Post-Workout Cards)	
Fri	☐ **Movement Prep Card** (pg 11)	h:　m:
	☐ **PFT** (pg 29)	h:　m:
	☐ **300 m swim** (pg 23)	h:　m:
	☐ **11 min tread** (pg 25)	h:　m:
	☐ **4 min float** (pg 25)	h:　m:
	☐ **Post-Workout Card** (pg 49)	h:　m:
Sat	☐ **Movement Prep Card** (pg 11)	h:　m:
	☐ **7-mile ruck, 1:55 or less** (pg 26)	h:　m:
	☐ **Assess feet/gear** (pg 51)	h:　m:
	☐ **Post-Workout Card** (pg 49)	h:　m:

WEEK ❻ Notes (additional exercises, nutrition, sleep, fatigue)

Sun

Mon

| Energy level | 10 | 9 | 8 | 7 | 6 | 5 | 4 | 3 | 2 | 1 |

Tue

| Energy level | 10 | 9 | 8 | 7 | 6 | 5 | 4 | 3 | 2 | 1 |

Wed

| Energy level | 10 | 9 | 8 | 7 | 6 | 5 | 4 | 3 | 2 | 1 |

Thu

| Energy level | 10 | 9 | 8 | 7 | 6 | 5 | 4 | 3 | 2 | 1 |

Fri

| Energy level | 10 | 9 | 8 | 7 | 6 | 5 | 4 | 3 | 2 | 1 |

Sat

| Energy level | 10 | 9 | 8 | 7 | 6 | 5 | 4 | 3 | 2 | 1 |

WEEK ❼	⏱ Time
Sun	**Rest, stretch, hydrate & recover** (Movement Prep and Post-Workout Cards)

Mon		h:	m:
	☐ **Movement Prep Card** (pg 11)	h:	m:
	☐ **1.5-mile run** (pg 29)	h:	m:
	☐ **Movement Card #1** (pg 30)	h:	m:
	☐ **1.5-mile run** (pg 29)	h:	m:
	☐ **Post-Workout Card** (pg 49)	h:	m:
Tue	☐ **Movement Prep Card** (pg 11)	h:	m:
	☐ **1-2 mile ruck, faster than 13 minutes per mile** (pg 26)	h:	m:
	☐ **Movement Card #2** (pg 31)	h:	m:
	☐ **1-2 mile ruck, faster than 13 minutes per mile** (pg 26)	h:	m:
	☐ **Post-Workout Card** (pg 49)	h:	m:
Wed	**Rest, stretch, hydrate & recover** (Movement Prep and Post-Workout Cards)		
Thu	☐ **Movement Prep Card** (pg 11)	h:	m:
	☐ **Swim Card #4** (pg 40)	h:	m:
	☐ **Post-Workout Card** (pg 49)	h:	m:
Fri	☐ **Movement Prep Card** (pg 11)	h:	m:
	☐ **10 x 200 m run** (pg 29)	h:	m:
	☐ **8 x 100 m swim** (pg 23)	h:	m:
	☐ **Post-Workout Card** (pg 49)	h:	m:
Sat	☐ **Movement Prep Card** (pg 11)	h:	m:
	☐ **8-mile ruck, 2:15 or less** (pg 26)	h:	m:
	☐ **Assess feet/gear** (pg 51)	h:	m:
	☐ **Post-Workout Card** (pg 49)	h:	m:

WEEK ❼ Notes (additional exercises, nutrition, sleep, fatigue)

Sun

Mon

| Energy level | 10 | 9 | 8 | 7 | 6 | 5 | 4 | 3 | 2 | 1 |

Tue

| Energy level | 10 | 9 | 8 | 7 | 6 | 5 | 4 | 3 | 2 | 1 |

Wed

| Energy level | 10 | 9 | 8 | 7 | 6 | 5 | 4 | 3 | 2 | 1 |

Thu

| Energy level | 10 | 9 | 8 | 7 | 6 | 5 | 4 | 3 | 2 | 1 |

Fri

| Energy level | 10 | 9 | 8 | 7 | 6 | 5 | 4 | 3 | 2 | 1 |

Sat

| Energy level | 10 | 9 | 8 | 7 | 6 | 5 | 4 | 3 | 2 | 1 |

WEEK ❽	⏱ Time	
Sun	**Rest, stretch, hydrate & recover** (Movement Prep and Post-Workout Cards)	
Mon	☐ **Movement Prep Card** (pg 11)	h: m:
	☐ **3-mile run** (pg 29)	h: m:
	☐ **Movement Card #3** (pg 39)	h: m:
	☐ **Post-Workout Card** (pg 49)	h: m:
Tue	☐ **Movement Prep Card** (pg 11)	h: m:
	☐ **1-2 mile ruck, faster than 13 minutes per mile** (pg 26)	h: m:
	☐ **Movement Card #4** (pg 40)	h: m:
	☐ **1-2 mile ruck, faster than 13 minutes per mile** (pg 26)	h: m:
	☐ **Post-Workout Card** (pg 49)	h: m:
Wed	**Rest, stretch, hydrate & recover** (Movement Prep and Post-Workout Cards)	
Thu	**Rest, stretch, hydrate & recover** (Movement Prep and Post-Workout Cards)	
Fri	☐ **Movement Prep Card** (pg 11)	h: m:
	☐ **PFT** (pg 29)	h: m:
	☐ **300 m swim** (pg 23)	h: m:
	☐ **11 min tread** (pg 25)	h: m:
	☐ **4 min float** (pg 25)	h: m:
	☐ **Post-Workout Card** (pg 49)	h: m:
Sat	☐ **Movement Prep Card** (pg 11)	h: m:
	☐ **9-mile ruck, 2:55 or less** (pg 26)	h: m:
	☐ **Assess feet/gear** (pg 51)	h: m:
	☐ **Post-Workout Card** (pg 49)	h: m:

WEEK ❽ Notes (additional exercises, nutrition, sleep, fatigue)

Sun

Mon

| Energy level | 10 | 9 | 8 | 7 | 6 | 5 | 4 | 3 | 2 | 1 |

Tue

| Energy level | 10 | 9 | 8 | 7 | 6 | 5 | 4 | 3 | 2 | 1 |

Wed

| Energy level | 10 | 9 | 8 | 7 | 6 | 5 | 4 | 3 | 2 | 1 |

Thu

| Energy level | 10 | 9 | 8 | 7 | 6 | 5 | 4 | 3 | 2 | 1 |

Fri

| Energy level | 10 | 9 | 8 | 7 | 6 | 5 | 4 | 3 | 2 | 1 |

Sat

| Energy level | 10 | 9 | 8 | 7 | 6 | 5 | 4 | 3 | 2 | 1 |

WEEK ⑨	⏱ Time

Sun	**Rest, stretch, hydrate & recover** (Movement Prep and Post-Workout Cards)	
Mon	☐ **Movement Prep Card** (pg 11)	h: m:
	☐ **1.5-mile run** (pg 29)	h: m:
	☐ **Movement Card #5** (pg 35)	h: m:
	☐ **1.5-mile run** (pg 29)	h: m:
	☐ **Post-Workout Card** (pg 49)	h: m:
Tue	☐ **Movement Prep Card** (pg 11)	h: m:
	☐ **1-2 mile ruck, faster than 13 minutes per mile** (pg 26)	h: m:
	☐ **Movement Card #6** (pg 36)	h: m:
	☐ **1-2 mile ruck, faster than 13 minutes per mile** (pg 26)	h: m:
	☐ **Post-Workout Card** (pg 49)	h: m:
Wed	**Rest, stretch, hydrate & recover** (Movement Prep and Post-Workout Cards)	
Thu	☐ **Movement Prep Card** (pg 11)	h: m:
	☐ **Swim Card #5** (pg 41)	h: m:
	☐ **Post-Workout Card** (pg 49)	h: m:
Fri	☐ **Movement Prep Card** (pg 11)	h: m:
	☐ **6 x 400 m run** (pg 29)	h: m:
	☐ **3 x 300 m swim** (pg 23)	h: m:
	☐ **Post-Workout Card** (pg 49)	h: m:
Sat	☐ **Movement Prep Card** (pg 11)	h: m:
	☐ **10-mile ruck, 2:30 or less** (pg 26)	h: m:
	☐ **Assess feet/gear** (pg 51)	h: m:
	☐ **Post-Workout Card** (pg 49)	h: m:

WEEK ⑨ Notes (additional exercises, nutrition, sleep, fatigue)

Sun

Mon

| Energy level | 10 | 9 | 8 | 7 | 6 | 5 | 4 | 3 | 2 | 1 |

Tue

| Energy level | 10 | 9 | 8 | 7 | 6 | 5 | 4 | 3 | 2 | 1 |

Wed

| Energy level | 10 | 9 | 8 | 7 | 6 | 5 | 4 | 3 | 2 | 1 |

Thu

| Energy level | 10 | 9 | 8 | 7 | 6 | 5 | 4 | 3 | 2 | 1 |

Fri

| Energy level | 10 | 9 | 8 | 7 | 6 | 5 | 4 | 3 | 2 | 1 |

Sat

| Energy level | 10 | 9 | 8 | 7 | 6 | 5 | 4 | 3 | 2 | 1 |

WEEK ⑩	⏱ Time
Sun	**Rest, stretch, hydrate & recover** (Movement Prep and Post-Workout Cards)

Mon		
	☐ **Movement Prep Card** (pg 11)	h:　　m:
	☐ **1.5-mile run** (pg 29)	h:　　m:
	☐ **Movement Card #1** (pg 30)	h:　　m:
	☐ **1.5-mile run** (pg 29)	h:　　m:
	☐ **Post-Workout Card** (pg 49)	h:　　m:
Tue	☐ **Movement Prep Card** (pg 11)	h:　　m:
	☐ **1-2 mile ruck, faster than 13 minutes per mile** (pg 26)	h:　　m:
	☐ **Movement Card #2** (pg 31)	h:　　m:
	☐ **1-2 mile ruck, faster than 13 minutes per mile** (pg 26)	h:　　m:
	☐ **Post-Workout Card** (pg 49)	h:　　m:
Wed	**Rest, stretch, hydrate & recover** (Movement Prep and Post-Workout Cards)	
Thu	**Rest, stretch, hydrate & recover** (Movement Prep and Post-Workout Cards)	
Fri	☐ **Movement Prep Card** (pg 11)	h:　　m:
	☐ **PFT** (pg 29)	h:　　m:
	☐ **300 m swim** (pg 23)	h:　　m:
	☐ **11 min tread** (pg 25)	h:　　m:
	☐ **Post-Workout Card** (pg 49)	h:　　m:
Sat	☐ **Movement Prep Card** (pg 11)	h:　　m:
	☐ **12-mile ruck, 3:00 or less** (pg 26)	h:　　m:
	☐ **Assess feet/gear** (pg 51)	h:　　m:
	☐ **Post-Workout Card** (pg 49)	h:　　m:

WEEK ⑩ Notes (additional exercises, nutrition, sleep, fatigue)

Sun

Mon

| Energy level | 10 | 9 | 8 | 7 | 6 | 5 | 4 | 3 | 2 | 1 |

Tue

| Energy level | 10 | 9 | 8 | 7 | 6 | 5 | 4 | 3 | 2 | 1 |

Wed

| Energy level | 10 | 9 | 8 | 7 | 6 | 5 | 4 | 3 | 2 | 1 |

Thu

| Energy level | 10 | 9 | 8 | 7 | 6 | 5 | 4 | 3 | 2 | 1 |

Fri

| Energy level | 10 | 9 | 8 | 7 | 6 | 5 | 4 | 3 | 2 | 1 |

Sat

| Energy level | 10 | 9 | 8 | 7 | 6 | 5 | 4 | 3 | 2 | 1 |

M✦RSOC.COM

MARCH 2018

www.ingramcontent.com/pod-product-compliance
Lightning Source LLC
Chambersburg PA
CBHW040908210326
41597CB00029B/5018